Quotes About Michael Toms

". . . one of the best interviewers who has ever worked the American airwaves, radio or TV."

Robert Fuller
Physicist, educator, past president of Oberlin College, and active citizen diplomat

"Someone with whom I have cruised some important realms of the cosmic ocean and in doing so have developed ever increasing confidence in his intuitive navigation."

R. Buckminster Fuller (1895–1983)
Inventor of the geodesic dome, designer, philosopher, and creator of the World Game

". . . Bill Moyers and Michael Toms are alike: two of the most creative interviewers it has been my good fortune to work with."

Joseph Campbell (1904–1987)
Mythologist and author of *Hero with a Thousand Faces, The Masks of God, Myths to Live By,* and *The Mythic Image*

New Dimensions Books

Fritjof Capra
in Conversation with
Michael Toms

Series editor
Hal Zina Bennett, Ph.D.

Published by

Aslan Publishing
Lower Lake, California
USA

Published by
Aslan Publishing
P.O. Box 108
Lower Lake, CA 95457
(707) 995-1861

For a free catalog of all our titles,
or to order more copies of this book
please call (800) 275-2606

Library of Congress Cataloging-in-Publication Data
Capra, Fritjof.
    Fritjof Capra in conversation with Michael Toms.
       p. c. — (New dimensions books)
    ISBN 0-944031-40-4 : $8.95
    1. Capra, Fritjof. 2. Religion and science. 3. East and West.
4. Spiritual life. 5. Human ecology. I. Toms, Michael.
II. Title. III. Series.
BL240.2.C274 1993
291.4—dc20   93-37976
                                                                           CIP

Copyright © 1994, New Dimensions Foundation

All rights reserved. No part of this publication may be reproduced, stored in a retrieval system or transmitted in any form or by any means, electronic, mechanical, photocopy, recording or otherwise, without prior written permission from Aslan Publishing, Lower Lake, California, USA, with the exception of short excerpts used with acknowledgment of publisher and author.

Cover design by Channing Rudd
Printed in USA

10 9 8 7 6 5 4 3 2 1

Table of Contents

Series Introduction 7

Introduction 11

Section One
The Marriage of Modern Science and Spirituality 15

Section Two
The Global Impact of Modern Science 41

Biographical Notes on Michael Toms 83

About New Dimensions 85

Series Introduction

Through a cooperative arrangement between New Dimensions Radio and Aslan Publishing, we are pleased to present these highly readable introductions to some of the finest and most influential minds of our times. These books grew out of interviews by Michael Toms, of the "New Dimensions" radio series, a man who is arguably the most articulate, insightful and well-read radio personality of our times. A non-profit educational organization, New Dimensions Radio has dedicated its efforts to "foster communication about personal and social transformation" through weekly interviews with some of the world's most prominent figures in religion, science, psychology, philosophy, ecology and virtually every other discipline, for over twenty years.

In 1973, Michael and Justine Toms were inspired by a comment made by Charles Tart, a renowned researcher in altered states of consciousness. Tart had remarked that we were "in the midst of the most dramatic shift in human consciousness in the history of the planet—and nobody is paying attention." Convinced that Tart was right, and that the mass media were particularly resistant to what was happening, Michael and Justine stepped boldly forward and launched "New Dimensions," originally through KQED-FM in San Francisco. Michael has served as principal host, with Justine acting as program and series producer.

Today, the "New Dimensions" radio series boasts a broadcast network of nearly two hundred stations throughout the United States and abroad. Together, Michael and Justine have pioneered a new style of "compassionate journalism," which fosters open dialogue and clear communication intended to empower the listener. Their weekly listenership has now reached more than two million worldwide, while the roster of their interviewees includes Nobel Prize physicists, holistic physicians, artists, healers, and notables such as Buckminster Fuller, Joseph Campbell, Black Elk and the Dalai Lama.

Our goal in this series is to offer books that extend the Tomses' ambitious efforts to the printed page, probing the minds of our country's leading thinkers, recording their thoughts and feelings as well as the essence of their own works. In these books, we feel that we have been able to capture more than the intellectual side of the great minds of our generation; in addition, we've been able to catch glimpses of these great minds as people, touched by those events of everyday life that we all know so well. These are books that allow us to share how the most outstanding minds incorporate their ideas into their own daily life processes.

By virtue of the more informal interviews, which are the basis of these books, we catch brief glimpses into the doubts, fears and private aspirations that interviewees have expressed. These candid insights are rarely found in more formal books or lectures. They are important because they help us humanize works that we might otherwise think are beyond our own capacities. In this way, the *great works* of our time become much more accessible, much more "user friendly" and applicable in our own lives.

At the time the publisher came to me with the idea for this series, I had been an avid listener of New Dimensions radio programming since its inception. I had even enjoyed the honor of being interviewed by the Tomses over the years, talking about books I had either authored or co-authored. But most important of all, the interviews I heard on that series were often my introduction to the real giants at the cutting edge of science, psychology, the social sciences, spirituality, the arts and self-development.

From the beginning, we have all been excited by the opportunity to present in this highly readable form ideas that are helping to make this a better world. It is our hope that whether you are reading these ideas for the first time, or have picked up this book to refresh your memory of a lecture, book or workshop that previously introduced you to these authors, lecturers, and teachers, that you will be moved as profoundly as we have been by their efforts.

Throughout our publishing program, we continue to dedicate ourselves to bringing you the works of men and women who are amongst the most creative, thought-provoking, and controversial on this planet. We hope you will look for other books in this series, since together they present the ideas that are truly changing the world for the better.

—Hal Zina Bennett

Introduction

In this book, we present the work of Dr. Fritjof Capra, whose vision has provided us with a view of our world and our lives which touches all of us, even those who have never previously read his books or heard his lectures. We believe that this introduction to his work will help to bring about a new appreciation not only for Dr. Capra's contribution but for the global awareness he has spent his life exploring and teaching.

Fritjof Capra was raised in the European academic tradition, taking his doctorate at the University of Vienna in 1966. He then went on to do research in theoretical high-energy physics at the University of Paris, the University of California at Santa Cruz, Stanford University, and Imperial College in London.

In 1975, Shambhala Publications published his book *The Tao of Physics: An Exploration of the Parallels Between Modern Physics and Eastern Mysticism*. With the publication of that book, Dr. Capra set in motion a dialogue that would revolutionize the way we all look at the world in which we live. For the first time in history, we began to see that ancient eastern philosophers, utilizing intuitive processes only, had observed and articulated a picture of the universe that we would begin to grasp with our most advanced scientific methods only thousands of years later.

Dr. Capra's work has provided us with clarity about the profound interactive nature of the world in which we live. The vision he has provided, through his books and lectures, has taught us to think globally, to see the major problems of our time as interconnected and interdependent—that is, systemic rather than localized. It has led to an awareness of *deep ecology*, a way of looking at life that is earth-centered, respecting the diversity and richness of all life forms.

Today, Dr. Capra continues his work through the Elmwood Institute Center for Ecoliteracy, in Berkeley, California. At this ecological think tank, one finds leaders from all areas of interest, from Nobel Prize physicists to psychologists and politicians, participating in the ongoing dialogue to better understand the world in which we live and expand ecological literacy.

Programs at Elmwood include regular series such as "Deep Ecology Dialogues," which help provide a basic understanding of deep ecology and how it can be applied in all disciplines and in all areas of our personal lives. The goal is to ultimately see ecological literacy become the central organizing principle in education and business, globally as well as locally.

If you are interested in learning more about Elmwood's principal work, further information is available in the book *EcoManagement*, published by Berrett-Koehler. Other books by Dr. Capra are *The Turning Point* and *Uncommon Wisdom*. He also co-wrote the screenplay *Mindwalk*, a feature film based on his work.

Further information about the Institute can be obtained by writing to: **The Elmwood Institute Center for Ecoliteracy,** 2522 San Pablo Avenue, Berkeley, California 94702.

Section One

The Marriage of Modern Science and Spirituality

MICHAEL: I'd like to go back to the late sixties. You had recently come to California and you met J. Krishnamurti. Could we talk a little about that period of your life?

FRITJOF: Well, it was at this time that I encountered the eastern spiritual traditions. I came to California in the fall of '68, from Paris, where I had spent my first two years as a post doctoral researcher in physics. In Paris I had already met people and read books about eastern traditions. I read about Zen, the *Bhagavadgita*, Hinduism and Buddhism. I read Alan Watts' book *The Way of Zen* and some of his other works. But I had not met any spiritual teachers yet. When I came to California I went to the University of California at Santa Cruz, which was then a relatively new campus and pretty far out, as we used to say then. There were some very unconventional people on the faculty and even more unconventional lecturers were invited. And Krishnamurti was one of them. He came to Santa Cruz for three days and gave three consecutive lectures. What he said deeply affected me. He did not refer to any particular spiritual tradition but just to spirituality as such. He addressed the subject in a very direct, very original way. What most challenged me was that he focused on the importance of going beyond thinking, of being present and communicating in a different way, beyond time and beyond thought. He impressed me at the same time that he disturbed me very much.

I was a young theorist just beginning to make a career of thinking...paid for thinking. Now here is this eastern teacher, who was already an old man, to me looking very much like Castaneda's Don Juan. The Castaneda books were being published about that period of time and so I thought of this man, this Krishnamurti, as my own Don Juan, sitting there telling me to stop thinking. So, you can imagine, it really disturbed me. I tried to talk to him after his lectures but it was very difficult to arrange a meeting with him. Finally I succeeded, however, and the conversation we had was really one of the first major turning points in my life; he solved the dilemma I was having in one sentence, you know, at a stroke, like a Zen master. And I asked him, as a scientist just starting my career, how could I follow his advice and stop thinking and "achieving" as he called it; how could I attain what he called "freedom from the known." He looked at me and said, "first you are a human being, then you are a scientist. As a human being you can reach your wholeness only through meditation." We talked about what he meant by that and then he told me that once I had reached that centered state then I could truly think.

Krishnamurti said that he thought science was wonderful. There was no question of my giving up science. "I love science," he said. "But that's only part of you and the whole is much more." And that solved my problem.

I think this is a very significant story, not only for me personally but also in terms of our whole culture. I think that most scientists are threatened by this comparison between the ideas of modern physics and the ideas of the mystical traditions that I present in my books, because they see the teachings of mysticism as an attack on their science. In the parlance of the scientific community, to call something mystical is to say that it is highly suspect and very unscientific. And so, to tell a physicist that his cherished theory is very similar to a theory in the mystical tradition is a deep insult. It's very threatening. This became very clear to me in the reactions to my book that I got from the scientific community. It helps to remember that I was once threatened in very much the same way.

MICHAEL: In your book *The Tao of Physics,* you said that scientists do not deal with truth; they deal with limited and approximate descriptions of reality. Perhaps you could go further with that.

FRITJOF: Well, this is something extremely important in modern science and I should clarify what I mean when I say "truth." I mean it in the sense of seeking an exact correspondence between the description of the thing and the actual thing itself. That kind of truth does not exist in science. That kind of truth does not exist in any kind of spoken or written word. Our language and our rational minds are just not capable of achieving that kind of truth. This is the concept that Lao Tse was reflecting on in the *Tao te Ching,* when he said that those who speak do not know those who do not speak. You cannot express the full truth in words, you can only approximate it.

That awareness has come to the fore more and more in science during this century. In this respect, I've been very deeply influenced by my physics mentor, Geoffrey Chew at the University of California, Berkeley, with whom I worked for over ten years; he places great emphasis on "approximation" in science.

MICHAEL: Can you explain what you mean by "approximation" as it applies to science?

FRITJOF: Yes, of course. Let me describe it this way. Chew was once interviewed by the BBC and at the end of the interview they asked him what he saw as the next big breakthrough in science. They expected him to tell them about a grand unified theory or a new description of gravity or an exploration of outer space. But, instead, he told them that the greatest breakthrough would be in the recognition that whatever we say is limited and approximate. It was Werner Heisenberg who in a sense discovered the concept of approximation in atomic physics. He said that in quantum theory every word, every theory that we might propose, clear as it might seem to be, is only an approximation.

MICHAEL: Well, how would you contrast, say, Geoffrey Chew's work with, say, someone like Stephen Hawking?

FRITJOF: Well, first of all, we should say that they are both at the same level, at a genius level that is very rare. And I think that Steve Hawking would agree with Chew that everything we say is approximate. I don't know Hawking well enough, philosophically, you know, to be able to speak with any authority about him. But in science today there are few good scientists who would not agree with that statement, though they would often act as if a certain theory was the truth and would not question it. By contrast, scientists like Chew and Hawking are ready to question their own theories continuously.

MICHAEL: Hawking says that in ten to twenty years computers will be able to discover the secret of the universe. And physicists will no longer be necessary because computers will be able to spit out the formulas. Now I don't suppose that Geoffrey Chew would agree with that.

FRITJOF: No. Not at all. I think this is very much tied in with Hawking's personal tragedy—that he is paralyzed and so the separation of mind and body, which he has practiced in a brilliant and very moving way for so many years of his life with a lot of positive effects, allows him to separate his mind from his body as no other scientist does. He needs computers to assist him because he has lost his ability to speak normally. So I think it's only natural that he has gone in this direction, to see computers as a sort of scientific panacea. I understand his position, but it's definitely not a position that I share.

MICHAEL: Why would you say that Geoffrey Chew's path in physics is a non-traditional one?

FRITJOF: It is non-traditional because he questions the notion of fundamental entities. You see, in western science, we have for centuries expressed our knowledge using the metaphor of a building. We talk about knowledge standing on firm foundations, and by that we mean fundamental equations, fundamental constants, fundamental particles, fundamental principles. We talk about building blocks, the basic building blocks of matter. These are all architectural metaphors. In Descartes' conception of his new science, he starts with the statement, "I think, therefore I exist." Then he builds on that. He states that he examined the knowledge of his time and found it standing on shifting foundations, on sand and mud. And now he, Descartes, was going to build a new kind of science on firm foundations. Three hundred years later, when Albert Einstein talked about the beginnings of quantum physics, he used exactly the same language. Einstein said, "I could not find any firm ground on which to build." It was as if the ground had been pulled out from under him. I think Chew is the first who does not need firm ground, because he has replaced the metaphor of the building with the metaphor of the network. He sees knowledge as an intricate network of concepts, ideas and—in science—experiments and observations. There is no primary, no secondary, no up or down. There are no foundations. That's radical, that's really radical.

MICHAEL: As I hear you talking about Geoffrey Chew's work, I am reminded of David Bohm's "Implicate Order." Could you comment on that?

FRITJOF: I have always thought that their work was quite similar. In fact, many years ago the similarities in their work brought them together. They had known each other, of course, because they were both part of the same physics community, but they had not shared much direct exchange. I invited Bohm to come to the Lawrence Lab in Berkeley for a seminar with Chew, and that was very stimulating for both sides.

To show you the similarities in their thinking let me talk a little more about this network structure, both of our knowledge of that structure and of what we observe in physics.

Ever since Heisenberg, Niels Bohr, Schroedinger and other quantum physicists, we have known that the physical reality at the atomic level does not appear as a collection of separate objects—building blocks or anything like that. Rather, we see it as a network of relationships or inter-related processes. If we try to imagine this interconnected network, which Bohm also calls the "unbroken wholeness," and we continue to simultaneously talk about elementary particles and electrons and atoms and molecules, we must inevitably bump up against the question, how can we talk about these so-called separate objects if there are no such things as objects? Here we must realize that from within this unbroken interconnected network we are able to pick out certain patterns. In order to observe them and define them we cut them out of the network, both conceptually and physically. This is part of our experimentation and observation.

So imagine this interconnected network. Then imagine that in certain parts we see certain patterns. We decide that in order for us to study them we have to cut those patterns out by cutting through the interconnections. And you can do that in several ways—when you see the network there is no dotted line that says cut here, right? So you have to decide for yourself, as the scientist, as the experimenter, where to cut. So a certain subjectivity has to come into this game. You as the researcher are imposing something of yourself on the experiment at this point.

Now we come to what is perhaps Heisenberg's greatest discovery, that we can never talk about atomic or subatomic phenomena without talking also about ourselves, about our experiences and our involvement in the reality we are observing. Both Chew and Bohm have continued with this line of reasoning. Both have emphasized that ultimately, if we are to move forward in science, we will have to understand consciousness, the nature of our consciousness. We'll have to understand that part of consciousness that makes the decisions about where to cut and why. Until we understand consciousness, we cannot understand physical reality.

So this is one area where Chew and Bohm are very similar. But there are other areas, of a more formal nature, where they think alike. They agree, for example, that the ideal mathematical language for describing a network structure is topology. Both also use matrices, another mathematical tool, when describing this interconnected structure. So there are great similarities. As you know, I predicted in *The Turning Point* that the two approaches will eventually converge and coalesce.

MICHAEL: You know, I was just thinking about David Bohm and the time he spent, over a period of twenty years or so, in dialogue with Krishnamurti, exploring the nature of consciousness. And I'm reminded of the story you wrote in *Uncommon Wisdom* about Geoffrey Chew and his son. Maybe you could tell us that story?

FRITJOF: Yes, of course. His son was a student of Mahayana Buddhism and at this particular time Chew was preparing for a trip to India, where he was scheduled to do some lectures. His son, who was maybe eighteen or twenty at the time, told him that before he went to India he should know that his boot strap theory was very similar to this mystical theory of Mahayana Buddhism. And Chew was absolutely shocked. He had the same position that I described when I first met Krishnamurti, about being extremely suspicious of anything mystical. But, like me, he changed completely. The discovery that his son had been right was a major turning point for him. He confessed that after his initial shock he was deeply moved, experiencing a sense of great wonder and being grateful to be living at a time when these connections were being discovered.

MICHAEL: So here are these two physicists who have resonated with spiritual tradition and have also seen the parallels in their own work.

FRITJOF: Right, and this is actually how I started *The Tao of Physics*—by showing that the leading physicists of our century have sought inspiration from the east. As they confronted the radically new concepts of nuclear physics, they became very disconcerted and were just at a loss to explain things in traditional terms. They looked around in philosophy and religion, and wherever else they could find help, and many of them looked towards the east. It wasn't just a handful of renegade scientists that did this. It includes people like Heisenberg and Niels Bohr and Robert Oppenheimer, Erwin Schroedinger, Wolfgang Pauli—they all made contact with eastern traditions, and not only superficial contact. For instance, Robert Oppenheimer learned Sanskrit so that he could read the *Bhagavadgita* in the original. Schroedinger went deeply into the Upanishads. Bohr was so extremely impressed by Chinese wisdom that he made a special trip to China.

The leading physicists of our time turned to the east and found valuable help that they apparently could find nowhere else; you might say that they had to look to the east because they could find no other system of thought that helped them interpret what they had begun to see as a whole world view. It's important to understand that they were not influenced by the east so much as they found in eastern thought a way to interpret and understand the scientific implications which no other thought system adequately touched.

MICHAEL: You yourself spent some time at the Esalen Institute, that mecca of consciousness, as it were. One of the people you met there was Gregory Bateson. How important was he in your own development?

FRITJOF: Bateson was a major influence in my thinking—and in my profession. At the time I met Bateson, I was preparing my second book, *The Turning Point*, and had begun reading about biology, psychology and social issues. But I approached these from the vantage point of a physicist. I had this idea that Newton and Descartes had influenced the entire range of classic science, not only in physics and chemistry but also in biology, psychology, economics and other social sciences. All these sciences modeled themselves after the model established in Newtonian physics. This was even true of Freud's theories of psychoanalysis. And so I thought, now we have a new physics and I can show where it takes us—I can interpret this new science for people who are not physicists. I felt I could hold up the new physics as a shining model for a new kind of psychology, economics and so on. And this was a great mistake because I fell into the same Cartesian trap—and Gregory Bateson was the one who saved me from that error by pushing me in a different direction, forcing me to put life at the center of all my considerations. He did this in a way very similar to what happened for me with Krishnamurti. However, Bateson didn't tell me directly to my face; he said it to a friend. Somebody asked him what he thought about my work. And he said, Capra, the man is crazy! He thinks we're all electrons! When this was reported back to me I was really shocked, but I knew that he was right. It was, of course, vastly exaggerated, but he really put his finger on it. After that, I had many conversations with him. He really changed my thinking by showing me that the change of paradigms in physics is very important as an historical example—but it could not serve as a model for similar changes in other sciences.

MICHAEL: What was your personal impression of Gregory Bateson?

FRITJOF: Well, I was sort of overawed by him, especially in the beginning. He was a giant in every respect. He was very tall, very big, a very tall man. And he was a giant intellectually. He also had this very typical British way, a British academic way of speaking, where he would make allusions to other thinkers or other theories and concepts, and much of the time you would have to guess what he meant. He didn't spell it out fully. He had this dry, English humor. And so a conversation with Bateson was always a little like shadow boxing. I was quite intimidated, as were many others. It took me a long time…in fact, I hardly ever had a casual conversation with him. I always felt I had to sort of seize on the moment and either prove myself or take advantage of the time I had with him and ask him something important. I did have a number of very significant conversations with him. After a couple of years I got so I could just sit down for a cup of coffee with him. But that took a long time.

MICHAEL: In your book *Uncommon Wisdom* you told about a conversation you had with Gregory Bateson in which he spoke of metaphor as the foundation of life. What exactly did he mean by that?

FRITJOF: Right. I think this shows how difficult it was to understand the man. I mean, you ask yourself what the hell does he mean when he says that metaphor is at the fabric of life. What he meant was that the way to understand life is to penetrate into that network structure, the fabric. This is very similar to what Bohm and Chew were saying. To understand that interconnected network Bateson talked about "patterns." Once somebody asked him what is a pattern, and he replied that it was anything you saw twice. Which is also a very great answer.

Bateson talked a lot about patterns. One of his famous phrases was "the pattern which connects." He meant that when you look at two life forms, two plants or two animals, you see similarities; these similarities are not because they belong to the same class, though that also may be true, but because there are similarities in their dynamic patterns. He used logic and a sort of anti-logic to illustrate this point. Let me try to capture the way he did this.

Bateson would write a syllogism on a blackboard, one of the classical Greek constructs, such as: Men die. Socrates is a man. Therefore, Socrates will die. Next to that, Bateson would write: Men die. Grass dies. Therefore, men are grass. Now this is something totally different, challenging the expected logic. However, the similarity in the syllogism is in the dying, not in the class of objects or subjects. So Bateson says there is a certain pattern reflected in this syllogism, and the pattern in this case is the dying. You observe that pattern in humans and you observe it in all living things; therefore, men are grass.

For Bateson, then, *men are grass* is the metaphor, reflecting on the way nature is built, the living nature, and it is also the way poets think. And this is why sometimes poetic language is more appropriate for describing this aspect of nature. I would like to just tell you a personal story which also illustrates this same point, in a way that perhaps is easier to grasp.

When my daughter was two-and-a-half years old, we watched a video of an ice skating competition. One of the performers was a woman who looked like my daughter's babysitter, Michike. My daughter, Juliette, said, "That's Michike!" We watched that tape very often and Juliette loved to watch it. Every time she saw this particular performer, she said, "Oh, there's Michike." I began wondering if she really thought that this was her babysitter Michike or if she used the name metaphorically because this was somebody who looked like Michike. She didn't explain it that way, of course, she just said "That's Michike." Then one day she asked, "Daddy, what's Michike's name?" Then it was clear to me that Michike had been her metaphor. I told her the name of the skater and from that time on she didn't call her Michike any more. It was a very nice example of how children use metaphors in this way. I observe that with her very often. Sometimes, when she saw a little girl in a storybook, she'd say, "That's Juliette!" She didn't mean that she was in the book. She meant only that it was a little girl like her. And so that's how Bateson used metaphor and that's, of course, how poets use metaphor.

MICHAEL: I can't help but reflect that many great religious documents are also written in metaphor.

FRITJOF: Yes. And people like Joseph Campbell and others have made the point that poetic language and myth are in fact the closest we can get to truth. That metaphorical sense that we evoke when we say men die, grass dies, therefore men are grass—at one level this is nonsensical. At another level it's profound. If I asked you what is profound about it, you wouldn't be able to say right away. But there is definitely something in that syllogism that strongly resonates with us, as it does in all poetry and art. And so this level of metaphorical description is much deeper and much more resonant than the straight logical language.

MICHAEL: Isn't it interesting that in this technological age of ours "myth" implies that it's unreal or false.

FRITJOF: Yes, and you know, Kumara Swami, who was a great scholar of Indian mythology, said that myth is the closest to truth that you can get.

MICHAEL: While you were at Esalen, you also did some work with Stan Grof and had some interesting experiences.

FRITJOF: Yes. I have known Stan Grof since the mid-seventies and we have been close friends. I worked a lot with him.

I got to know him very well personally. I also explored many things with him professionally and joined his seminars and workshops. He attended seminars I gave and I participated in his, and we did things jointly. Our friendship began when I visited him first in his house at Big Sur, a very beautiful house on a cliff above the Pacific Ocean. And he was kind enough to let me use his house several times when he traveled. He travels a lot and gives a lot of workshops in Europe and all over the world. For weeks at a time I would stay in his house and those were among the happiest and most centered moments in my life. I would go there to work and my entire day became a day of work and meditation. I remember that I constructed a sun dial because I didn't want to use clocks. I wanted to really live out there on the cliff over the ocean, with the sun coming up in the morning. During a very intense period of work I would start in one corner of his big room and be fully dressed because it was cold in the morning. As the sun came up I would move my table across the room to stay in the shade, shedding my clothes along the way, and then I'd be sitting practically naked in the hot sun, putting on my clothes again as the sun went down. And this was a cyclical rhythm that went on for several weeks. I also fasted during that time so it was just an extreme experience of work and meditation which was really beautiful.

MICHAEL: It's interesting that both you and Stan Grof are European and both of you found your way to California. Do you have any comment on your own personal journey that way?

FRITJOF: Well, I think it may be typical with people who are developing this new way of thinking, this new paradigm. Gregory Bateson was, of course, European. So was Alan Watts. There are many others who have lived in the States, or even in California, who have come from Europe. The advantage that we have is that we can see the European traditions and the new ideas from the outside and from the inside. So we have a broader perspective than either those who were raised in the States or those who stayed in Europe. In the social arena, for instance, many of us who came from Europe are very critical of the American way of life—the way society is organized, the way we use our natural resources, the way we have a sort of imperialist attitude vis-a-vis the Third World. Many of these attitudes are quite obvious to a European but not so obvious to an American. On the other hand, when you go to Europe you see that the people there are often seriously confined by their traditions, you know; they cannot broaden their consciousness because they are so boxed in. Living in California, then going back overseas, you can also see the European's shortcomings from the outside, even as you still experience it as a European because that's what you are on the inside. It's greatly enriching to be between two cultures like that. But I should tell you that it also has its price; you lose your native language, your mother tongue. Mine is German, of course. In a sense I speak English better than I speak German. But in another sense I also speak German better than English. It depends on what I am talking about. But I don't speak either German or English really perfectly, in the sense of having either language really under my skin. You lose that. For instance, I could never write poetry, given where I am with language. It's typical for poets and writers, literary writers, that they keep their native language. They might move to California but they might speak German and write in German. I write in English, so that's a price to pay but it does have great advantages.

MICHAEL: How would you say that Stan Grof's work has influenced your own?

FRITJOF: Well, as you know I have left the realm of physics and have explored psychology, psychotherapy, social issues and many other things. Stan Grof's work has helped me enormously to explore dimensions of consciousness from a scientific point of view, as well as from the point of view that combines first-hand experience and science. He is one of the very few people who is beginning to develop a new kind of science which talks about consciousness in scientific terms. He makes the point that you cannot measure and quantify consciousness but you can analyze patterns of experience. So instead of measuring and quantifying as you do in physics, you would share patterns of experience. That would be the data. And on the basis of shared patterns of experience you can create a model. You can analyze patterns and interconnect them, and that is what Stan has been doing. For me that has been extremely exciting to see—that somebody can go so far away from traditional, let's say Freudian, analysis, which was his field originally, and yet stay within the scientific framework.

MICHAEL: Yes, he was actually trained in Eastern Europe, and it's amazing that he was able to come so far from that kind of background and training, from both a Communist country and a very rigid kind of scientific training.

FRITJOF: He was also lucky, you see, because his job in Prague was to investigate LSD, which was only beginning to be researched. Originally it was believed that this drug induced a temporary state of neurosis or psychosis that could be studied as a model of those dysfunctions. But Grof discovered that LSD does not bring about mental illness but rather is a catalyst for looking deep into your psyche. That's really the starting point for his work.

MICHAEL: In the early seventies, you returned to Europe. While in England you met E. F. Schumacher. Perhaps you could tell us about that meeting and Schumacher's influence in your life.

FRITJOF: Well, I heard about Schumacher when I was writing *The Tao of Physics*. I was riding the underground and reading the newspaper when a headline about Buddhist economics jumped out at me. Since the book I was writing was essentially on Buddhist physics, that headline really stirred me. I read the article and it told about these economists on the British Coal Board who had gone to the east and were now introducing Buddhist values into western economics. They were questioning the whole value structure of western economics in this book called *Small Is Beautiful*. I read this review of Schumacher's book, put it aside and didn't follow it up right away because I was busy writing *The Tao of Physics*. However, after two years of living in California I had been deeply affected by the new social movements and had come to the conviction that our system of economics is bankrupt and we need to replace it with a more relevant system. I was aware of the ecology movement and the impact of conventional economic theory on our environment. So here, in Schumacher's little book, was somebody who had put this out in a very articulate way. But some years passed before I actually read his book. I was just very much impressed by it! And so it was natural for me when I wrote *The Turning Point*, some time after that, that I would think to contact Schumacher. He very kindly said that I could visit him in England, where he had a place in the country. So I went over and spent one afternoon with him...a remarkable afternoon.

MICHAEL: And then your interests in economics led you to Hazel Henderson, whom they call an uncommon economist.

FRITJOF: Yes, and that was also sort of miraculous, because in speaking with Schumacher I had found we had many things in common but also had a great many differences. In the end, I felt that his beliefs were really rooted in a patriarchal value system. I went away from our meeting thinking that what I wanted to find was an alternative economist with all the credentials of Schumacher, and all the wisdom and philosophy of Schumacher, who also would have the feminist consciousness. Naturally then, I thought, it would have to be a woman. And at times I gave up. I thought, that's a fantasy. But, lo and behold, I met this woman and that's Hazel Henderson, who at that time was not so well known as she is now. She is unique because, first of all, she is not an academic. She's an intellectual but also an activist. Even more than an activist she's completely self-taught, has established her reputation and her authority in various circles in Washington, in the corporate world, in the academic world. In addition, she has just a fantastic ability to explore new ideas on her own and formulate them in a very concise and very witty way. In many ways she is really one of the most remarkable people I have ever met.

MICHAEL: Soon after you finished *The Turning Point*, you got the idea of starting the Elmwood Institute. You mentioned somewhere that this was an outgrowth of that book?

FRITJOF: Yes. I had met so many outstanding people while writing this book and I felt that if I wanted to go on exploring these ideas I had to formalize the process in some way. So I founded the Elmwood Institute, in 1984, with Hazel Henderson and Stan Grof as members, along with various others. We now have a very exciting mix of theorists and activists. We explore new ways of thinking by bringing theorists and activists together. And so, you see, the work continues, exploring all the various ways that the new paradigm can shape our lives in the near future.

MICHAEL: I'd like to shift, for a moment, to your meeting with Indira Gandhi, which you talk about in the later part of *Uncommon Wisdom*. Perhaps you could tell us about that.

FRITJOF: I went to India at about the time *The Turning Point* was just coming out. I had an advance copy with me on the plane and the publishing date was going to be during the time I was in India. I ended up giving my only advance copy to Indira Gandhi. Symbolic gestures are important to me, and I remembered how the first copy of *The Tao of Physics* got to India in the pocket of Ali Akbar Khan, whom I had recently met. He had said, well, I'm going to India tomorrow and I will take it with me. To me this was very meaningful. So, the first copy of *The Turning Point* I took to India and gave to Indira Gandhi.

First of all I should say that I was very surprised and somewhat shocked by the celebrity I had in India, in the east in general, and in Japan. It has taken me a long time to understand it, but finally I understood that the part of my work which is so controversial here in the States, namely the eastern mysticism, is mainstream over there. They find the mysticism part of my writing very familiar and they are deeply interested in the modern science part of it. So I am well received in India by the establishment. You know, I gave an interview to the *Times of India*, I think it is called, and was on the front page of their newspaper. I was received by University presidents and people high up in government and finally by Indira Gandhi, with whom I had a very unusual and totally surprising one hour of conversation.

MICHAEL: Why was it surprising?

FRITJOF: It was surprising because I had a preconception of Indira Gandhi as this hard-driving, very powerful, very ruthless woman who knew how to play the political game and was very authoritarian. Instead, I met a woman who was very soft, very wise, very self-effacing, very—how should I say this?—she did not seem to have any ego at all. She was the exact opposite of what I had thought her to be, a preconception that I had learned from the media. I'm not saying that the other Indira Gandhi did not exist—the powerful politician and all that. And actually, during my visit, I once got an inkling of that. But she also had this other soft, feminine side which the media never showed. So the conversation I had is, I think, a rare document in itself.

MICHAEL: So what does the future hold for Fritjof Capra?

FRITJOF: Well, raising a young child has been a new and totally different kind of experience for me, and one that is very important. Beyond that, I'm going to continue building up the Elmwood Institute with my friends and colleagues. I'm going to continue to be an activist as well as a thinker and I have a long range plan, which I've had for some time now, to write another book, a very different book. It will be for children. It will be a book about—as Gregory Bateson said—"what it's all about." But it will be for ten-year-olds.

MICHAEL: Great! So all of us will be able to read it!

Section Two

The Global Impact of Modern Science

MICHAEL: Your book *The Turning Point* is as much philosophy as it is science. In it, you point to a major trend in our society's evolution. What I would like to know is what prompted you, a physicist, to write about society. It's not something that we generally associate with physics, is it?

FRITJOF: I have always been interested in the broader implications of modern physics, such as how our discoveries in this science impact on other systems of thought, particularly those which have a direct effect on society. This interest began when I was a student in Vienna. I was reading Heisenberg's *Physics and Philosophy,* in which he deals with the difficulties physicists had at the beginning of the century, especially in the 1920s, when they were just beginning to formulate their new ideas in atomic physics. He points out that what the new physicists were saying constituted a whole new world vision, making it necessary to revise any vision of the world that was based on Newtonian physics. For a long time there did not even seem to be a common language between those who still held to the Newtonian world view and those who were beginning to integrate the new world view ushered in by quantum physics.

So we begin to appreciate how the discoveries in quantum physics had a major impact not only on science but on how we would view the world from this time on. Heisenberg's point was that radical new concepts in science cause us, inevitably, to revise the ways we look at and relate to the world.

I read Heisenberg's work with great interest when I was seventeen or eighteen years old, and I've been interested in this subject of science's impact on society—and society's resistance to these new ideas—ever since. This same interest led me to begin exploring how the new world view that has been emerging for the past seventy years or so is influencing thought not only in my field of research, but in virtually all other sciences and systems of thought.

This interest in how science affects us prompted me to write *The Tao of Physics*, which relates the views of modern physics to the views of ancient mystics from the East. Since then, I branched out to look at the effects of physics on other fields—biology, medicine, psychology, anthropology and so on. It is obvious to me that we can no longer stay within a narrowly defined discipline, because the problems we are facing in society are systemic problems. They transcend any single discipline.

To even begin to address these systemic problems, we must move beyond the isolation created by any single specialization. But to accomplish this, we must create a common language to communicate our ideas. To give you an example of why this is important, I spent some time with Rupert Sheldrake, an English biologist with some very new ideas. He is a specialist in plant physiology, botany and biology and we wanted to converse at a quite sophisticated level. At the same time, we found that we couldn't use the technical jargon to which we were accustomed in our specialties. We had to express the essence of the various ideas in our fields in a language that the other could understand. So we ended up spending a good deal of our time and energy searching for ways to, in effect, translate our knowledge back and forth between the specialized languages of our disciplines.

And so I think if we are to go beyond the narrow confines of specialized disciplines, we need to be able to capture the essence of our sciences and relate them to a broader context. The way I see physics relating to society right now is rather indirect. In fact, this brings up a point that reviewers have made concerning *The Turning Point*. A few have said, well, here's Capra and he is saying that in light of the recent discoveries in quantum physics we have to structure society in a certain way. In a way, they're criticizing a physicist because, as they see it, I'm trying to be a sociologist. But I assure you that this is not at all what I'm saying.

I relate our social problems to physics by analogy rather than through cause and effect. My main analogy has to do with what physicists went through in the 1920s while they were trying to understand atomic and subatomic phenomena. The exploration of those phenomena led them to recognize the severe limitations of their specialized language as well as their previously held concepts, and their way of thinking and conducting experiments. They literally had to change their whole way of thinking and find entirely new concepts and research processes to deal with this new reality. The crux of the problem was that they were trying to apply an outdated conceptual framework—namely the framework associated with classical or Newtonian physics—to a new kind of reality. The problems they had with this threw those physicists into a very serious crisis. At first it appeared to be only an intellectual crisis, but it soon became apparent that it was much more than this. It was an emotional crisis and existential crisis as well. Eventually they worked their way through this crisis, as we well know by now, and were rewarded by deep insights into the nature of matter and its relationship to human consciousness.

It is my belief that we can learn a very useful lesson from these physicists' experiences, one that is applicable to all areas of society. And this is really the thrust of my work, that so many of the problems we face today are essentially crises of perception, just as was true for physicists in the 1920s. I am suggesting that our social institutions are organized around an outdated world view—the same outdated world view as the physicists struggled with, the world view of Cartesian/Newtonian science.

The challenges that were thrust upon us by discoveries in physics in the 1920s literally forced us to create a brand new world view, a new way of looking at our lives and the world around us. This is what is often called a paradigm shift, a shift in the way we view the world. Similarly, the challenges of our day are forcing us toward a paradigm shift in society. This is the parallel I see between what happened in physics in the 1920s and what is happening in society today.

MICHAEL: In writing *The Turning Point* you sought out experts from different fields of endeavor and talked with them about some of these ideas and what impact they have on society. Could you talk a little bit about how you did that?

FRITJOF: Yes. I think this was quite unique. I needed the other experts because in all my work I require a high standard of accuracy and expertise and I just knew I couldn't write about biology or economics or psychology, and these various fields, without consulting with leading experts in these fields.

And so I started looking around. But the way I found these people was rather remarkable, because I really just sort of came across them. Ever since I got in touch with the eastern philosophies, especially Taoism, which teaches us to let things come to us, I have trusted this process. Castaneda writes about this also. His teacher, Don Juan, talks about the "cubic centimeter of chance" that pops up every now and then. And he says that when you get your life in order—in his terms, when you become a warrior or man of knowledge—then you will quickly recognize this cubic centimeter of chance and pick up on it when it presents itself to you. It was like that for me. Somehow, my advisors sort of came to me or I came across them in this way. Perhaps together we created and recognized the cubic centimeter of chance. Let me give you a few examples.

One of the people who came into my life is Margaret Lock, a medical anthropologist who graduated from the University of California, Berkeley. She teaches now at McGill University in Canada and she's an expert on Chinese medicine. She wrote a book about the use of Chinese medicine in modern Japan, which is very valuable because it shows how a traditional medical system can be used in a modern context—and we can learn a lot from that. Well, Dr. Lock came to one of my lectures on "S-matrix theory," which is the field of physics I worked in. I lectured and drew diagrams on the board and at the end Dr. Lock came up and introduced herself and said, you know your diagrams look a little like acupuncture diagrams. And so we got into this very exciting conversation about the Chinese world view and how it relates to the views of modern physics. So we became friends and started working together.

Another person I met in this way was Stanislav Grof, about whom we spoke earlier. I attended a seminar he gave, in which he said that what we really need in psychology is a kind of "bootstrap approach." This is a term borrowed from my field of physics. It means that you have a network or mosaic of models that are interlinking, without having one overall theory that binds them together. They overlap, connecting one with another so that the places where they connect are virtually seamless. So you end up with a mutually consistent network of models, all supporting one another.

When I first met Stan Grof, he was advocating this approach for psychology and psychotherapy. As you might imagine, I was very excited by that and began talking to him and soon we started working together.

Another of my advisors was Carl Simonton, the cancer specialist who has done so much with the mind-body approach to cancer. Another was Hazel Henderson.

There was Gregory Bateson also. Though he was not officially a close advisor, he was very influential. All these people I met more or less by chance, but I don't like to use the word "chance" because there is more behind it than that. I asked each of them to write position papers on various topics, on various fields, such as economics, psychology, biology and so on. I used these in the book. Carl Simonton didn't want to write anything, but he came to my house and we had a non-stop conversation for three days, which I taped.

I assembled all these different parts together and we met again for four days in a house down at Big Sur. This was one of the most exciting things that I have ever done, because those people were all very close in their thinking, even though they didn't know each other. There were about ten people in all. We found that we were all expressing very similar ideas, and many of us had never met before. There was great excitement, sparks were flying between us. We discussed the whole framework of the book and then, after a period of time to integrate it all, I started writing. The book became a sort of grand synthesis of new ideas, showing how these ideas, all from very different disciplines, have many interconnections and interdependencies.

The participants in these conferences helped me throughout the writing of the book. I would often call them up and tell them, listen, I have this kind of problem, what do

I do with it? How does this fit in with this or that world view? They were extremely helpful and, of course, this is all described in the book.

MICHAEL: As you were talking, it just brought to mind the importance of the interconnections we need to build between one profession and another, one discipline and another. I was reflecting on how so much of our society is over-specialized, with no perspective on what's happening outside their little territory. What you've done is a reminder for us to become "comprehensive-ists" rather than specialists.

FRITJOF: Yes, absolutely. You see, Michael, the problems we are facing as a society—economically, environmentally, in the field of health, as well as with problems of violence, crime, dysfunctional families, our fear of nuclear war—all of these are systemic problems. That means they are not isolated, one from another, but are all interconnected and interdependent, like the nervous systems of our own bodies. Focusing only on one aspect, without taking the rest into account, doesn't really help in the long run, and in fact can be extremely harmful. You have to treat the whole patient, you see, not just one single part, as if it was operating in total isolation. When we try to isolate and focus only on one part, we may cause one problem to go away only to create another. If you imagine an interconnected network of relationships—which in fact is the reality of the new paradigm—we see that there are physical, psychological, social, and environmental relationships, and there are social systems embedded throughout, all of which makes for a very complex multidimensional network. If you don't take these interdependencies into account, then what you tend to do is shift problems from one area to another. I'd like to give you two examples, one from medicine and one from economics.

Let's begin with medicine. If we were to create a holistic or systemic view of health, we might view an illness in this way. Let's say it begins with a man living in a stressful situation. To escape the stress this man might take one or another escape route, such as alcohol, driving too fast, workaholism, and so on. These escape routes give the illusion of relief, perhaps, but they do not address the main issue. So the stress continues to accumulate, filtering through the mind-body system and then expressing itself in worsening symptoms which can be predominantly physical or predominantly psychological. But the truth is that you can't really separate the two, because physical and psychological components are present in all illness. Nevertheless, we can say that certain symptoms are predominately physical or predominately psychological. Or they can be predominately social. Then you have phenomena like anti-social behavior or disruptive behavior, excessive violence, crime, reckless driving accidents and so on.

Now let's say the person gets an infection at this point. Medicine comes to the rescue and treats the disease with antibiotics. You may get rid of the infection but you still haven't addressed the problem systemically because the stressful situation, which is the foundation of the illness, is still in place. So, in an effort to escape the stress once again, the person may go on to develop brand new symptoms, which may be psychological or social in nature. So you haven't solved the real problem at all. You've just shifted it around, you see.

Now a similar thing occurs in economics, for example, in what came to be known, in the 1970s and 80s, as "Reaganomics." There are many aspects to it, but a key aspect is the attempt to solve the problem of inflation by restricting credit. The Reagan economists did this by raising interest rates, making it more expensive to borrow money and therefore discouraging people from doing it. The theory was that there was too much money around and not enough being produced. So, the Reagan people theorized, you could solve the problem by restricting the money supply, the credit supply. Now, it is true that this brings down inflation a bit, but the cost of doing that is a tremendous increase in unemployment and the bankruptcy of a record number of businesses. So what is actually being accomplished? They are merely shifting from inflation to bankruptcy and unemployment. In the end, you haven't done anything to solve the problem. In fact, you may have made it much, much worse!

So these are two examples that demonstrate how the main problems of our times are systemic problems, all interconnected. What you have to do is change the structure of the system itself rather than focusing on a single problem in a narrow way. Each day it becomes increasingly important to have this broader view, to take into account how any single problem is actually embedded in a much bigger system and is related to many other problems. Only then will we be able to bring about positive and constructive change.

MICHAEL: We certainly can see these principles at work in the environmental crisis, which you also discuss in your book. You talk about the Love Canal, where an entire residential area was built over a toxic waste dump. It ended up seriously affecting the health of hundreds of people.

FRITJOF: Yes. You see the time scale is very important here. If you shift a problem onto the ecosystem, the ecosystem has capacity to absorb great masses of blows, over a long period of time, many more blows than, say, the social system can absorb. But, of course, the damage is being done, even if we don't notice it right away, and it comes back to us eventually, in ten years, twenty years, forty years, or in some cases the problems don't become obvious to us for hundreds of years.

So there are many different time scales. Problems are related not only in this network of interactions in the present but also to interactions in the future and the past. Or, take nuclear power, for example. While nuclear reactors may seem okay in the present they constantly emit radiation and are a grave danger to the environment and to future populations. Furthermore, it is not only the cumulative effect of radiation that we need to look at but also the storage, over perhaps hundreds of years, of nuclear waste from these power plants. So the short-sighted view that applies to the present situation without a projection of how these issues will affect us in the distant future, is very irresponsible and unwise. We need to remember that time must be factored into the picture if we are truly going to deal with things holistically, or systemically.

MICHAEL: But in today's technology, we don't seem to have the ability to think a year ahead, much less fifty or a hundred or a thousand!

FRITJOF: That's right, and it's a big problem. In technology, we are still looking at all these issues from a Newtonian perspective, that is, from a mechanistic world view. We are viewing the world as a machine—a reductionist approach, reducing things to smaller components, assuming that we can plan and control the whole universe through learning to control its smaller parts, one at a time. Specialization, compartmentalization, fragmentation of thinking—all of these belong to the Newtonian/Cartesian world view. No doubt we have to acknowledge that in the last three hundred years this world view has been very successful. It is not successful any longer, for a number of reasons. First, we live today in an overpopulated world where all phenomena are very closely interconnected. Everything happens faster and everything affects more people than ever before in history. Another reason is that we are rapidly running out of natural resources. If you live on a vast open continent, as the early American settlers did when they came here, with plenty of natural resources, the tendency is to be wasteful and you can function for a long time in an un-ecological way. And for a while it can appear that we are doing no harm. Now there are other native cultures, indigenous cultures like the American Indian, who had a very keen sense of the environment, of the ecological network, even when they were surrounded by great abundance. They expressed their reverence for the environment in their rituals and spiritual traditions.

While we may have once lived in a world of rich natural resources, where we could afford to be wasteful, we don't live in a such a world any longer. So this mechanistic, reductionist, Cartesian world view is now hopelessly outdated—though it is still a useful perspective for some issues. The broader management of our problems will have to be from a systemic, ecological world view. In a sense you could say that all problems we are facing right now are ultimately environmental problems, because they are all coming from the fundamental neglect of our interconnectedness, in our social organizations as well as in our individual lives.

MICHAEL: I remember when I was a teenager, in the 1950s. The future was painted bright and cheery. We were always hearing how atomic energy was going to deliver all kinds of wonders and there were so many people painting this bright and cheery future. But somehow that emphasis, which seemed to hold such promise for the future, has put us grossly out of touch with the earth. We thought technology was the answer, the solution, that it would enable us to gain control over nature. We'd dam up rivers, we'd overcome the dangers of natural disasters such as tornadoes and hurricanes. Science and technology would make the world safe for us all. We're all conditioned to think of the answers in mechanical solutions. Certainly there have been some benefits, but at what cost? Perhaps the biggest problem we face is one that we've created ourselves—being disconnected from Mother Earth.

FRITJOF: Yes, it's become a basic characteristic of our society to believe all the solutions can be found in technology. And that is really a direct consequence of our mechanistic world view. When you really believe the universe is a machine, no matter what the problem, it can appear that the best person to call in times of trouble is an engineer. We continue to pursue this approach in virtually every discipline. For instance, in defense, the question of national security is analyzed at the Pentagon and at the White House almost exclusively in terms of hardware; how many tanks and bombs do we need to be secure? They perceive a threat from another country and the answer is always the same. Invest in more hardware, get a bigger military budget. They completely overlook the fact that warfare is no longer a solution. Today's military has such sophisticated technology that if you brought it all into play there would be no winners. This whole question of warfare, which is central to the military, has become hopelessly outmoded because paramount to every war is the possibility of nuclear weapons being used, ultimately decimating our entire planet. How can we change all this? Only by shifting the whole framework, the whole paradigm, from the mechanistic to a systemic world view.

MICHAEL: It's also a good example of the absence of human values that has spilled over into the Defense Department. We don't even consider human values. We come up with terms like *mega deaths*...creating a way to think about killing that makes it seem that no real humans are involved. Just one more example of how we have to shift our perspective and get away from this mechanistic way of approaching our lives.

FRITJOF: And we have to say, of course, that this change is already beginning to occur, however slowly. When I was writing *The Turning Point* I remember wanting to be critical of the Cartesian paradigm, wanting to focus on it and analyze its structure and conceptual framework and show its limitations. I wanted to be as critical as I could be and also be accurate. And very often I overlooked the fact that change was already occurring. I think in the finished version of the book I took all this into account. But in the earlier versions of the manuscript, many people, my brother for one, read the manuscript and told me, you know, you're a little too critical here. These changes are already happening.

MICHAEL: Sometimes appearances can be deceiving. Outwardly, the world can appear to be crumbling but underneath some substantive positive changes are occurring.

FRITJOF: You see this when you adopt a dynamic perspective. This dynamic perspective is one of the main themes of natural ecology. It's an intrinsic aspect of nature...that in nature everything is always changing and dynamically interrelating, moving, transforming. And so when you look at the world with this dynamic perspective you see that cultures rise and fall. They have their own cycles, their own rhythms. The culture will climb to a certain pinnacle. Then, in an effort to maintain its present status, there develops a certain rigidity and the culture cannot change any longer, cannot respond to the current challenges. And while one culture is declining, a new one is already rising from the ashes. This is the rising culture I refer to in the subtitle of the book.

When people say that society is crumbling and everything is breaking down, what they are actually looking at is the old system breaking down and the new one rising.

I think it's important to look back and recall the changes one has seen in one's own lifetime. Although it's somewhat arbitrary to say this happened in the 50s and that happened in the 60s, it's a nice way of just parceling things out a little bit. In a way, you could say the 60s ended with John Lennon's death, which really occurred in the 80s. But nevertheless I think it can be useful to group things together in a certain way. Like you, Michael, I was a teenager in the 50s. James Dean was our hero at that time. He typified the youth culture of the 50s—he was literally the rebel without a cause. In the 50s, there was the friction between the generations, just as there always is when you have a youth culture, a culture of peers. The young people just get to the point that they don't believe their parents any longer. However, in spite of the friction James Dean's generation really shared the same world view as their parents, the same belief in technology, in progress, in gadgets, in the educational system. All this was never questioned in the 50s.

It was only in the 60s that the rebels began to have a cause, if you will. In the 60s there were two main movements, both in Europe and in the States. One was a social movement and one was a spiritual movement. The expansion of consciousness that occurred in the 60s, which was a tremendous adventure for all of us who went through it, went in two directions. One was toward a new kind of spirituality, the great interest in eastern culture and mysticism, in meditation and so on; the second direction was an expansion of political awareness, the questioning of authority. Throughout the 60s, the movement was toward a questioning of authority and toward the search for another world view. In the civil rights movement, the authority of a certain ruling class was questioned; in the free speech movement, the authority of the university was questioned. The same was true in France in May '68. I happened to be in Paris at that time and went through this whole student revolution of May '68. It was a questioning of authority, first in the university and then extending to society as a whole.

Then you had the emergence of a new women's movement, questioning the authority of men to determine the values of the culture. You had a revolution in humanistic psychology and in the health movement, questioning the authority of doctors and advocating the right of people to take responsibility for their own health. This whole questioning of the status quo was a strong theme.

Along with all of these challenges of the old world view was the expansion of consciousness toward spiritual values, what psychologists now call transpersonal experiences. I remember very well my feelings in the 60s when I was a hippie and we were questioning everything in society. We were living according to different values; we had different rituals, different clothes, a different style of living. But we couldn't express exactly what we were after…it was more of an intuitive feeling that was still developing.

I had discussions with my father about economics in the 60s and I tried to impress on him the fact that our whole economic system wasn't working any longer and we had to change it. But my arguments weren't very good. I had a very strong gut feeling that something was basically wrong with our system, but I didn't really know how to put it across. I think in the 70s there was a consolidation and integration of these new views and possible alternatives.

I also believe that one of the main social and political forces in our lifetime has been the emergence of feminine values and awareness. But by and large I see the 70s as a period of integration. Suddenly the magic of the 60s was gone. The initial excitement was gone and it was time to consolidate our views, focusing, digesting, integrating. I wrote *The Tao of Physics* in the 70s, and then went around lecturing and having these discussions with various people, always trying to integrate the new world view. I also believe that *The Turning Point* is a product of the 70s, inspired by the 60s.

In the 80s we began carrying out the transformation. Through the 80's, and now into the 90s, we are in a period of new activity, with a broader sense of integration but not just conceptual integration. Throughout the 80s we have seen people working as consultants, for example, bringing many of the humanist values of the 60s into the workplace and into medicine and into politics. And in the 90s, we have a Vice President who declares: "We must make the rescue of the environment the central organizing principle for civilization."

MICHAEL: As you were talking I was thinking about James Dean's movie "Rebel Without a Cause." The important statement there was that young people really didn't have anything to grab hold of, yet there was this resistance or denial of authority without any serious alternative. In fact, as you pointed out, they were still holding on to the old paradigm even as they rebelled against it. It was sort of a feeling that something was drastically wrong but you couldn't put your finger on it. It wasn't until Vietnam that there was a galvanizing of the energy, a cause, as it were, that would provide a way of articulating what was wrong. I think that the search for greater self-awareness, which emerges strongly in the 70s, relates to the integration that you were talking about.

FRITJOF: Talking about Vietnam and the anti-war movement, I think it's very interesting that the current equivalent to the anti-war movement, the anti-nuclear movement, is not so much an anti-movement as a positive movement. I would almost call the anti-nuclear movement an expression of ecological awareness, a feminine, pro-life movement; it is pro-human beings, pro-whole Earth...you know, not just against a war, not just against nuclear power and nuclear weapons, but moving forward in a very positive way, embracing a broad spectrum of the world population.

MICHAEL: I think it's very interesting that the anti-nuclear movement is also bringing together people who were once on opposite sides of the fence. You can even find military professionals who are saying, we've gone too far, we have too many weapons. You find the extremes coming together around the issue of nuclear proliferation.

As we talk about nuclear weapons, it is impossible to overlook the fact that we can trace their conception back to Einstein and the theory of relativity. Isn't it ironic that out of that work we should develop this whole path of mass destruction, as represented by the bomb? We do have a choice, do we not? We can choose to use the ideas of an Einstein, or the ideas we see in quantum physics, in either a constructive or destructive way. And in the case of Einstein, we chose the destructive route.

FRITJOF: In the case of Einstein and nuclear weapons, there were many different factors involved, certainly. But one crucial factor is that scientists have traditionally separated ethical and moral values from science. And this goes back to Galileo. It's also a very important part of what I call the Cartesian paradigm. Descartes separated the mind from the body and developed this mechanistic vision of the world. Around the same time Galileo combined observation of nature and the formulation of theories with mathematics. To do so he suggested that scientists henceforth should really consider those phenomena together and use mathematics to quantify them. And ever since Galileo, quantification and measurement has been a very essential part of science. Phenomena that couldn't be quantified were simply excluded from the domain of science. We developed a whole way of thinking that if a phenomenon couldn't be quantified in this way we would pass it off as not being real. So values and experience and quality have been excluded from science and this has had disastrous consequences. When combined with specialization you can see how easy it can be for a scientist like myself to exclude values in my day-to-day work. I don't have to think about human values or ethics when I work on my equations. But I certainly have to think about values when I decide what kind of research to do and how to apply it. When I place day-to-day problem-solving within the broader paradigm, within the broader context, values become absolutely crucial.

When we start separating values from scientific and technological activity, suddenly there is nobody to say, well, what are we doing with this new research? How are we going to use it?

I have observed that the most intuitive physicists, like Einstein and Niels Bohr, also have these strong social concerns. Because they are intuitive, they operate in a different mode of consciousness; they are able to perceive the wider context of problems. There are so many physicists like this now, scientists who have very strong social concerns, peace activists and so on, people who can see the disastrous consequences, who know where the research is leading and who speak out to warn us.

MICHAEL: Certainly in his later years Einstein was a powerful voice in speaking out against nuclear weapons.

FRITJOF: And he was not alone. Today we have to say that this myth of a value-free science is no longer possible. Scientists have to be responsible for their research, not only intellectually but morally. This is especially true in the case of nuclear weapons and nuclear power. Modern physics can be used in a very positive way, and this is what I am trying to do in my life...actually applying physics to show that it is in great harmony with similar approaches in other fields which give us this systemic, ecological vision. It is a vision that is ultimately spiritual—because the fundamental interdependence of all phenomena, the way in which the individual is embedded in the larger whole, is also a central issue in spiritual traditions. In this respect we have a marriage between modern science and spirituality. Once you have made contact in this way, and have adopted the ecological vision, your scientific research will lead you in very different, and I think positive, directions.

MICHAEL: One of the things that becomes clear as you talk is how the practice of science without values has impacted on other disciplines and other approaches. There is the tendency to say, let's remove the values so that we can be more "scientific." Let's just take economics, for example, where statistics can filter out human issues and concerns. We have practically made a religion of the so-called "bottom line," putting it ahead of values like quality of life.

FRITJOF: This is an excellent point. Economics was very strongly influenced by the conceptual framework of Cartesian-Newtonian science, and that's true for all the other sciences, not just economics. The economists took physics as a model because physics was sort of the pinnacle of sciences, the symbol of hard, exact science. Economists did two things which were in agreement with the model of classical physics.

For one thing, they applied reductionist thinking. They reduced the economy to various basic sets of phenomena. In doing so they separated the issues of economics from the social context in which it is embedded. They redefined basic concepts such as productivity, efficiency, gross national product, and so on, in a very narrow sense. They simply rejected the broader social and environmental context. Since they did this, their models became more and more unrealistic until now they are no longer adequate for mapping out or projecting future trends. The factors they are using simply don't take into account all the influences on economics that we find in an interdependent world.

The other thing economists did was to follow Galileo and his proposal of quantifying everything and rejecting whatever couldn't be quantified. This gave us econometrics, the quantitative approach to economics. They tried to build mathematical models that left out values and quality and human factors such as these. To use an analogy, what they have done might be compared to predicting the safety and expected lifespan of a new car by leaving out the fact that it was going to be operated by human beings.

Economics, of all the sciences, should be quintessentially concerned with human values and psychology because it involves production, distribution and consumption of goods, all of which depend on very complex human responses. What you produce and what you consume depends very much on your value system and on things like quality of life.

It seems obvious to say that if people are feeling economically insecure, they aren't going to purchase goods, and they aren't going to produce, and the economy is going to stagnate and crumble. That's the irony of going only by the bottom line, of putting profits first and simply leaving out unquantifiable human concerns. Of all the social sciences, economics should be deeply involved with the psychology of people. Some of the economists actually do that but it's all outdated. Economists haven't even gotten as far as Freud, let alone Maslow and humanistic psychology. So there is really a huge gap to be bridged.

While economists may pride themselves in basing their work on physics, we really need to face the fact that their model is Newtonian, which is a very, very old model, as science goes. They almost assume that economic activities take place in a vacuum. You know, they discuss market forces as if these forces occurred without any friction. You see, Newtonian mechanics neglects friction because when you deal with balls rolling down planes, pendulum swinging, and so on—and this is really how primitive that model is—friction is not very important. It's generally neglected. But in economics, in the social realm, there are very complex social frictions involved, environmental frictions which are, at best, extremely challenging. There are social and environmental consequences associated with every economic activity—factors that are usually grossly neglected. Such factors are called "externalities," meaning that they are outside the reach of present economic models. And obviously, since these social and environmental costs are now the main source of inflation, it's not surprising that even today's top economists can't figure out how to manage inflation. They can't do so because in the interest of modeling themselves on what they perceive to be a "scientific" mold—ancient Newtonian concepts—they have rejected key factors which either drive or depress an economy!

MICHAEL: So a practical example of that might be a nuclear plant that was planned solely from the standpoint of productivity, or how much money it could make for the utility company in the immediate future, but with no economic or social considerations given to such things as disposing of dangerous waste, having to dismantle the plant in the distant future, or the long-term cost to the environment.

FRITJOF: Right, the health cost and the social cost in terms of the sense of security or the fear that people would have, you know, the protests that might occur and so on. And if you were to factor in all these costs, it would make the cost of constructing a nuclear power plant completely out of reach.

MICHAEL: In the case of Three Mile Island, we had the very real possibility of a major utility company going bankrupt. There was also a federal court ruling that psychological values have to be taken into consideration when reopening a plant where there has been a nuclear accident.

FRITJOF: So these are all issues we must take into account in economics. Hazel Henderson, whom I mentioned earlier, says that "inflation can be defined as all the variables that economists have excluded from their models." That's a witty but precise way of pinpointing the problem. Another basic assumption of current economic theory is that undifferentiated growth is a sign of a healthy economy. We're talking about economic growth, technological growth and institutional growth. These are seen as economically healthy regardless of the frictions that take place, regardless of social breakdown, corporate crime, the threat of nuclear war, environmental disasters—all of that, which is directly or indirectly caused by undifferentiated economic growth. It's the idea that if we all simply maximize our profits, this will bring the greatest benefit to society. Underlying all this, of course, is the basic paradigm that the whole is nothing more than the sum of its parts. The idea that the interplay between the parts might either diminish the whole, or possibly expand it, is not entertained. This is what's going on in society now, for the most part, people maximizing their profits and generating a lot of social friction, which in turn requires a lot of litigation, leading to a lot of stress, alcoholism, social breakdowns of all kinds, health costs and so on. All these contribute not to a healthy economy or healthy society but, on the contrary, to inflation and social dysfunction. The current economic models, patterned after classical science, are just not able to take these complex human factors into account.

MICHAEL: I'd like to read something from your book *The Turning Point;* in it you say: "President Lyndon Johnson needed advice about warfare in Vietnam. His administration turned to theoretical physicists, not because they were specialists in the methods of electronic warfare but because they were considered the high priests of science, guardians of supreme knowledge. We can now say with hindsight that Johnson might have been much better served had he sought his advice from some of the poets, but that of course, was, and still is unthinkable."

Fritjof, in a recent conversation I had with Robert Bly, certainly one of America's leading contemporary poets, I read these same lines to him and asked him to comment. Here's what he said:

> ROBERT BLY: I can't imagine Johnson doing it—no! And I don't know, I mean I don't even like his statement. What do poets know? Why are poets put in this special category as opposite to nuclear scientists?
>
> MICHAEL: Let's go back to Plato when he talked about when philosophers shall be kings.
>
> ROBERT BLY: A dopey idea, a really dopey idea. I don't like Plato all the way down the line. I don't like the way he sneered at poets and kicked them out because they were too excitable and I don't like the idea that philosophers would be kings. I wouldn't want them as a king. And I'm not talking about smart ones like Jacob Needleman, I'm talking about the average run of philosophers. I'd rather be governed by a good Irish politician, to tell you the truth.

MICHAEL: There's Robert Bly presenting the poets.

FRITJOF: Well, you know, I used this Lyndon Johnson story to point out that if a President were to make decisions in a more holistic, systemic way, according to the new paradigm, he might better consult a Walt Whitman or a Gary Snyder, those ecological poets who have such a strong sense of the embeddedness of humanity in nature, in a larger whole, and who have a very strong sense of the value of human life. Certainly I count Robert Bly in this category of poets—that they would have a very useful role to play. Not in terms of diplomacy and detailed politics, but in terms of being advisors to the politicians to provide a broader vision. In our generation, poets like Bob Dylan and John Lennon, poets of rock and roll, offer a poetic vision that politicians would be well advised to heed. Take John Lennon's "Imagine" as a piece of poetry; it has a strong vision which is also political and it provides a very different way to look at the world. Interestingly, when John Lennon was shot politicians around the world commented about his passing, so in some small way at least his poetry had an impact on the political consciousness of our time.

MICHAEL: Well, in defense of Robert Bly I have to add that he often plays the role of the whimsical trickster. You can't always take what he says at face value. I continued our conversation that day and he came up with a very interesting reference to Heisenberg, the physicist. Here's how that conversation went:

> MICHAEL: Robert, what Capra was really trying to say is that there needs to be a deeper expression, a deeper contact in our political leadership, there needs to be an expression of that deeper part of ourselves of which you were talking earlier, and when one thinks about that, when one looks at the relationship between the arts and other events in our culture, one sees the art scene as being outside the rest of society. Somehow it's okay to

play war games at the Pentagon and then go listen to the symphony or go watch the latest Broadway play. Somehow the two things are separate. What's going on? What do you see that process as being?

ROBERT BLY: Well, that's very hard to get into. Many people in humanities have not had faith in their own deep tradition, going all the way back, and they will flip over and start praising and flattering scientists. And the scientists have not kept their connection with the deep roots of the human things. It's very interesting that science has produced certain breakthroughs in a human way, this moved me. Jacob Needleman mentioned this too. When he was about thirty or so, the ones who were asking for the new language were not the humanities people but the scientists, especially the nuclear ones, the subatomic ones. I heard Heisenberg the other day on the radio, speaking from Canada, and he said something like this: Well you know for many years we thought that light was a wave and what we understand now that somebody says well, it's actually little particles so we figure we are going to get a whole bunch of experiments and we're going to find out whether it's a wave or a particle, so now we've done all the experiments and the trouble is that's it's both a wave and a particle. So what does it mean? It means that our language...can't describe the world that we've got in front of us. So what are we going to do? We're going to have to make us a new language.

I loved that! That's exactly what is being asked for in the arts.

MICHAEL: Fritjof, any comment?

FRITJOF: Well you know this brings us back to the very beginning of our conversation because I told you I was very influenced by Heisenberg's book *Physics and Philosophy*, where he discusses this question of languages. I knew Heisenberg; I visited with him several times in Munich and had long discussions with him. That question of language and the conceptual framework was really at the very center of his thinking. That was the experience that physicists went through in the 1920s. When we turn to the present situation I agree very much with Robert Bly, that we need to make contact between the sciences and the humanities; in fact, the ecological vision I'm advocating goes way beyond science. You can formulate the ecological vision in terms of science, in terms of systems theory but you don't need to be a scientist to have the ecological vision. Actually, it helps if you're not, because the intuitive approach somehow is easier. That's what Robert Bly just said on this tape. We do have a long tradition of ecological visionaries like St. Francis of Assisi, for instance, whom you could call a patron saint of ecology. We have the poets, the transcendentalists; we have philosophers like Heidegger for instance, and Spinoza, the various philosophical and artistic roots in our culture...Then there are the Native American cultures that I've mentioned already, the various early cultures, which are highly ecological. We can make contact through those traditions and develop the framework which includes science but goes beyond it because it deals with a direct, intuitive awareness of the oneness of life, of its multiple changes and transformations, its cyclical paths of connections, its rhythms and patterns, an ecological awareness that is spiritual in its ultimate essence.

MICHAEL: You've mentioned a number of influences on your work and I know one of the things that's influenced you greatly has been the women's movement in America, the feminist movement. Perhaps you could talk a little bit about the influence that this has exerted on your own work.

FRITJOF: Well, I think that this influence actually goes back to my early childhood. I grew up in Austria on a farm during and right after World War II. My father and my two uncles were away in the war and the household and farm were run by women. After the war, when the men came back, the household was more or less still run by women. There were three women, my grandmother, my mother and my aunt, and I think they influenced my view of women and their role very strongly. I was used to women in positions of power and authority. I remember quite clearly that on the second floor of this farmhouse there was a balcony. After lunch my aunt would go out on that balcony and give various orders to people who worked on the farm. So I had this experience throughout my childhood, of the women in charge giving orders. I've never had a problem with women in positions of power and authority. In the 70s when I wrote *The Tao of Physics*, I lived in London. I read Germaine Greer and had friends who were in the women's movement, and was very much influenced by them. Feminism since then has been an increasing concern of mine. When I wrote *The Tao of Physics* I was careful to use non-sexist language. It's quite funny that when I was working on the second edition of *The Tao of Physics* I discovered in my first edition a lot of language that didn't seem sexist to me when I wrote it but now seems quite sexist to me. It reveals our own evolution.

The whole question of feminist awareness is very central to the new world view. First of all, when we talk about an imbalance of values in our society what we notice is that we have consistently favored one set of values over another. I use the Chinese terminology of yang and yin to say that we favored the yang values and neglected the yin counterparts. For example, we have overemphasized self-assertion and have neglected integration. We have favored competition and neglected cooperation; we have favored analysis and neglected synthesis; have emphasized expansion and exploitation and have neglected conservation, and so on. These values that the Chinese call yang, that are the favorite ones in our society, are also the values of patriarchal culture. The feminist movement points out that the Cartesian world view and this yang-oriented value system have been supported by the patriarchy. But like the Cartesian world view, patriarchy is now in its decline. It's a slow and painful decline but it is definitely progressing, and we are gradually seeing the feminist perspective becoming an essential part of the new vision of reality.

Another way to look at the feminine perspective is through the close connection between woman and nature. There have been several books that point this out. There's a book by Susan Griffin, a poet, entitled *Woman and Nature*. There's a book by Carolyn Merchant, an historian of science, called *The Death of Nature*. They address the theme in very different language but really make the same point. Then there is Charlene Spretnak's anthology, *The Politics of*

Women's Spirituality, addressing the whole broad framework of spirituality, ecology, feminism and politics. Now, what all these books show us is that woman and nature have been identified with one another throughout the ages. The exploitation of nature since the scientific revolution and the mechanistic world view of the seventeenth century have gone hand-in-hand with the exploitation of women by men. The whole language expressed man's domination of nature. So there is a natural kinship between the women's movement and the ecology movement. One of the very exciting developments of our time is the convergence of these two movements, with people actually talking about "eco-feminism" as the convergence between ecology and feminism. Also, you could say that the intuitive ecological perspective is closer to women. You know, women go through their biological cycles in a very pronounced way, and it is a constant reminder to them of their connection with a larger whole. Men also have cycles but we don't recognize them so much; our culture doesn't pay so much attention to the biological cycles that men go through. But certainly women know all that very well. The cycle of death and rebirth, the fact of giving birth, which is a direct experience for women and an indirect experience for men.

Many leaders in this rising culture that is working to bring together various disciplines and social movements are women. There's Hazel Henderson, Joanna Macy, Frances Moore Lappé, to name just a few. I think it's significant that they are women, because the women's movement and the feminist awareness can be seen as a catalyst to bring together the various social movements. This is what I see coming in future decades, the various social movements of the rising culture flowing together to become a powerful force for social change.

MICHAEL: When you bring up the feminist movement and its effect on you, I cannot help but wonder about how your work has been affected by your contact with eastern mysticism and the philosophies of the east. What about your spiritual life and how that has affected your work?

FRITJOF: Let me say first that for me Taoism comes closest to the modern scientific view. I am very essentially a scientist, you know, even when I talk about other things, or write about things which you have called philosophical. I write and talk about spirituality but my mind really works in the scientific sense. I have this fascination and joy in doing science and applying the scientific method to the analysis of various problems. But I'm not a narrow scientist. When I discovered the mystical or spiritual traditions, which came so very close to describing science, this was obviously just a path for me to follow. I practice Taoist techniques of meditation, tai ch'i and others, and my tai ch'i Master is also my doctor. He is an acupuncturist and herbalist, so this whole Taoist way of thinking has very strongly influenced my life, ever since the 60s. Castaneda exerted a strong influence on me too. I think his books come very close to describing a Taoist approach to life. Once I gave a lecture about the Tao of Don Juan, pointing out the very close parallels between Taoism and the way of the Yaqui shaman.

MICHAEL: How does the Yaqui shamanism, as expressed by Castaneda, relate to your book *The Tao of Physics*?

FRITJOF: Well, again, it is the ecological perspective emerging, the idea of being embedded in the natural environment and acting accordingly. The teachings of both Don Juan and Lao Tzu are aimed at showing us how our own actions are embedded in a larger whole and how you should flow with it, not try to force anything but flow with it. Don Juan talks about the way of the warrior, the way of the hunter, how to deal with decisions, how to deal with time, with personal history, many, many things that are really very Taoist in their approach.

MICHAEL: Besides Taoism and Yaqui shamanism, you have been deeply influenced by Buddhism. Perhaps you could talk about that.

FRITJOF: All these three, Taoism, Buddhism and the teachings of Don Juan, address themselves to the very essence of spirituality, the experience of reality, self-realization and so on. Over the last ten years or so, I have come to believe that it's not possible for us in the west to just follow purely eastern teachings. We have to adapt them to our own circumstances and our own cultural situations.

I've observed many teachers from the east, many gurus in America and in Europe, who absolutely could not understand the problems of our time, of our culture, and were not able to use the symbols of our culture to express the teachings of theirs. So I think we have to express the teachings in our own symbols. I have come to believe that ecology, in its deepest sense, may well be our western equivalent of an eastern mystical tradition. Buddhism, however, is adaptable to various cultural situations. It originated in India, then went to China and various Asian countries, ending up in Japan and finally jumping over the ocean to California.

As far as I'm concerned, I was most deeply influenced by Buddhism's emphasis on compassion and the achievement of knowledge. According to the Buddhist view there can be no wisdom without compassion. I have recently recognized that in a sense my first two books really follow the Buddhist approach. You could say that *The Tao of Physics* is concerned with wisdom and *The Turning Point* is concerned with compassion. *The Turning Point* is really motivated by social concern. In it, I point out the cultural crisis and then the cultural transformation, showing a way in which the transformation can be achieved, synthesizing various ideas from other people. But the whole book is really motivated by compassion. I have come to believe very strongly that this is what our science needs—compassion and wisdom. In a sense, Michael, we have been talking about compassion all along in this conversation—when we talked about values, the role of poets and politics, and so on. What we are really saying is that we need compassion in our politics, in our social interactions, in our science.

In science there is another figure, from the seventeenth century, whom we haven't mentioned. And this was Francis Bacon. Bacon dramatically changed the motivation of scientists and the whole approach toward knowledge by emphasizing domination and control. He developed the experimental method and the inductive method of science and was a fanatic experimenter. He said that nature had to be *forced* into a scientific experiment. He used very violent terms in connection with nature. And so, ever since Bacon, most scientists have cast themselves in this role of domination and control over nature.

Before Bacon came along, science was a matter of tuning in to nature and observing it without disrupting it. Medieval scientists said that they did their science for the glory of God. In the east the Taoists said that they flowed in the current of the Tao. Now these were all ecological and integrative in their approach, expressed in various terms, in various cultures. And the motivation behind science seemed to be an ecological and integrative one. At the core of this approach is intuitive and compassionate understanding, these being essential if we are to take human consciousness into account, which is why I think we must bring compassion back into science. For example, biomedical research that harms, tortures or kills animals is not the kind of science we need now, because it is devoid of compassion. Although some would argue that such research is compassionate because it ultimately benefits human life, what we gain in treating animals in this way we lose by giving up our compassion toward every living being. It is my feeling that the price is too dear to pay. We really must dramatically shift our motivation in science, shift the paradigm and world view and values that have dominated our thinking over the past few centuries.

MICHAEL: If someone came to you not knowing or being familiar with any of the ideas we've been expressing in this conversation, how would you describe *The Turning Point* to them?

FRITJOF: I would tell them it is a book about a fundamental change of the world view in science and in society. It is about a major cultural transformation. It is about changing the mechanistic world view described in the seventeenth century science of Descartes, Newton, Galileo, and Bacon. It is about changing the world view that sees the universe as a mechanical system, that separates the mind from the body and divides the world into separate entities made of fundamental building blocks, fundamental substances and so on. The new world view is a holistic and ecological one, emphasizing the interdependence of all phenomena and the intrinsically dynamic, ever-changing nature of reality. The basic thesis of the book is that the major problems of our time arise because our social institutions are still hanging on to the outdated Newtonian/Cartesian world view. The various crises that we have been observing for the past three hundred years, and are still observing, are all facets of this core issue—a crisis of perception.

The first part of the book reveals the historical development and limitations of the Cartesian paradigm in biology, medicine, psychology and economics. The second part presents the emerging world view, starting from physics, then going on to living organisms, and social systems. We also begin to explore how, with a more integrative approach, physics, biology, psychology and so on are themselves transformed, no longer specialized disciplines that isolate themselves from the larger context or system that is the world. It is about the movement toward a systems view of the world. This systems approach, which is consistent with the discoveries of modern physics, can be applied to the study of all living organisms, as well as the mind, human consciousness, evolution and complex social phenomena. All these are parts of the new vision of reality.

MICHAEL: You've spoken on these subjects internationally, and I cannot help but wonder how your ideas have been accepted, particularly by the media.

FRITJOF: There is media and there is media. The establishment media, of course, have been very critical, which you have to expect when you criticize the establishment. But I see cultural transformation taking place in the media themselves. There are media now that are definitely part of the rising culture. You yourself, Michael, with your New Dimensions Radio, are a driving force in this movement. I have acknowledged that again and again and been impressed by your work. As you know, you are not alone. There is a whole network of these media. Of course, I find them very open to my ideas because we have all gone through similar experiences and express similar ideas.

But my intent is not just to talk to people like you and me who are already convinced. I very much want to reach out to other segments of society. It's very important that we communicate with corporations, with schools, with the medical establishment and eventually even with the military.

MICHAEL: It seems extremely important to reach the large corporations and the military...the military-industrial complex.

FRITJOF: They are so often people who are trapped in the outdated world view, and it is, as you say, so important to reach them. We are all, as you know, struggling with this transformation from the old paradigm to the new. I don't want to give the impression that I have achieved this new realization and am perfect in my lifestyle and so on. I'm just struggling like everybody else. I'm very often Cartesian and sexist and racist and reductionist and so on. At this point, we certainly can't expect perfection, but much is accomplished by being aware of it and trying our best to move beyond the limitations of the old paradigm. I know there are already many people in the corporate world who feel this way, who share our views but may require support, and it is these people I would like to reach.

MICHAEL: In a very real sense the title of your book, *The Turning Point,* provides us with the image of our turning a corner. We haven't quite made the turn yet, and we really haven't yet seen what's around on the other side. We're in the process and we need to be open to the fact that as we make the turn we may see some things that we never expected to see.

FRITJOF: Of course, we are really building something new. The technologies we need must be ones we perhaps haven't yet even imagined. And it's not a question of going back in time, to borrow ideas from the Middle Ages, or some other time. We can't turn back the clock or solve our present problems in terms of what may have worked well enough in the past. We are in the process of discovering something new, something related to concepts and visions of past societies but also based on modern science and the new world view—but we don't know all the details. This is just a beginning.

MICHAEL: If you were to name the most important thing that we could do in the midst of this transition, what might that be?

FRITJOF: There's no single thing, but once you have recognized and adopted the ecological vision, then it is surprising how many things we can do. There are movements addressing themselves to various aspects of this new vision of reality: the ecology movement, the feminist movement, the antinuclear movement, the human potential movement, the holistic health movement, consumer movements, liberation movements of various ethnic minorities and so on. Some people may want to work within one of these. It almost doesn't matter which part of the paradigm you choose to work on since every part affects the whole. We can do many things in our daily lives, all of which are important, are making a contribution—recycling glass and paper, pushing your lawn mower instead of using power, driving a high-mileage car, riding a bicycle for short errands instead of taking the car, using public transportation, eating healthy, getting physical exercise, resolving our addictions, becoming more aware of sexist, racist, or exploitive language, participating more fully in the raising of our children, especially for us men—it's all very crucial. All of these things combine to lead us to this new vision of reality and cultural transformation. Everything we do, think and feel is important.

Biographical Notes on Michael Toms

Michael Toms is recognized as one of the leading spokespersons of "new paradigm" thinking. His perspective has been influenced greatly by his work with the late Joseph Campbell and Buckminster Fuller. He is perhaps best known as the host and executive producer of the widely acclaimed and award-winning "New Dimensions" national public radio interview series. He is Chairman Emeritus of the California Institute of Integral Studies, and currently serves as Senior Acquisitions Editor with HarperCollins San Francisco. His previous books include the bestselling *An Open Life: Joseph Campbell in Conversation with Michael Toms* and *At the Leading Edge: New Visions of Science, Spirituality and Society.* His interviews with leading thinkers of our time are the subject of our extensive "New Dimensions Books" series, edited by Hal Zina Bennett.

About New Dimensions

Inspired by the need for an overview of the dramatic cultural shifts and changing human values occurring on a planetary scale, New Dimensions Foundation was conceived and founded in March 1973, as a public, nonprofit educational organization. Shortly thereafter, New Dimensions Radio began producing programming for broadcast in northern California. Since then, more than 4,000 broadcast hours of programming intended to empower and enlighten have been produced. In 1980, "New Dimensions" went national via satellite as a weekly one-hour, in-depth interview series. More than 300 stations have aired the series since its inception, and "New Dimensions" has reached literally millions of listeners with its upbeat, practical, and provocative views of life and the human spirit.

Widely acclaimed as a unique and professional production, New Dimensions radio programming has featured hundreds of leading thinkers, creative artists, scientists and cultural and social innovators of our time in far-ranging dialogues covering the major issues of this era. The interviews from which this book was compiled are representative.

As interviewer and host, Michael Toms brings a broad background of knowledge and expertise to the "New Dimensions" microphone. His sensitive and engaging interviewing style as well as his own intellect and breadth of interest have been acclaimed by listeners and guests alike.

New Dimensions Radio provides a new model for exploring ideas in a spirit of open dialogue. Programs are produced to include the listener as an active participant as well as respecting the listener's intelligence and capacity for thoughtful choice. The programs are alive with dynamic spontaneity. "New Dimensions" programming celebrates life and the human spirit, while challenging the mind to open to fresh possibilities. We invite your participation with us in the ultimate human adventure—the quest for wisdom and the inexpressible.

For a free *New Dimensions Journal,* including a list of radio stations currently broadcasting the "New Dimensions" radio series, or an audio tape catalog, please write New Dimensions Radio, Dept. CB, P.O. Box 410510, San Francisco, CA 94141-0510; or you may telephone (415) 563-8899.

New Dimensions Tapes with Fritjof Capra

Physics and Reality with Fritjof Capra
The author of *The Tao of Physics* (Shambhala, 1975) takes us on a journey into the nether realms of quantum physics, where the traditional worlds of science and spirit are taking a new shape. Includes some remarkable commentary on future possibilities. A tape for modern thinkers.
Tape #1356 1 hr. $9.95

The Turning Point with Fritjof Capra
A renowned physicist and author of *The Tao of Physics* turns his attention to the sweeping social changes brought about by what he describes as "the rising culture." As in his book *The Turning Point* (Simon & Schuster, 1982), Dr. Capra addresses the state of the environment, modern industrial economy, and the choices individuals make in everyday life.
Tape #1652 1 hr. $9.95

The Coming Revolution with Fritjof Capra
"We think we understand how the world works, but we've got it all wrong," says Capra. He shows how the revolution in modern physics foreshadows an imminent revolution in all the sciences and a transformation of our world view and values.
Tape #1685 2 hrs. $15.00

Greener Tomorrows with Ernest Callenbach, Fritjof Capra and Hazel Henderson
Three of the cultural visionaries of our time come together to explore the emergence of a new paradigm for viewing reality based on the fundamental interdependence of all life with the cycles of nature. They stress the shift from a mechanistic and patriarchal world view to a holistic and ecological view, if we are to survive as a species and a planet. Callenbach, the ecotopian philosopher; Capra, the physicist and author of *The Tao of Physics* and *The Turning Point*; and Hazel Henderson, economist, futurist and author of *The Politics of the Solar Age* (Anchor Press, 1981) are founding council members of The Elmwood Institute, an organization created to foster new paradigm thinking.
Tape #1956 1 hr. $9.95

Quantum Odyssey with Fritjof Capra
Fritjof Capra relates his personal journey of meeting and interacting with some of the leading contemporary thinkers and visionaries. What emerges is a wisdom-packed dialogue about new ways of thinking and being. From Krishnamurti's spiritual insight to Gregory Bateson's intellectual prowess, Capra weaves an intricate and informative web of influences on his own pioneering work of bridging science and spirit. For anyone interested in the future, this is a conversation not to be missed.
Tape #2094 1 hr. $9.95

Mindwalk: The New Paradigm with Fritjof Capra
Capra describes the making of the 1991 film *Mindwalk,* based on the thought-provoking ideas presented in his book *The Turning Point.* From Descartes and Newton to Einstain and Bohr, he takes us on a journey through the past four centures of the scientific revolution and leads us to the birth of a new world view in response to escalating world crises. Capra reveals the difficulty in translating abstract ideas to the screen, and how the use of visual metaphor become a powerful tool and symbolized the importance of the ecological dimension.
Tape #2263 1 hr. $9.95

New Paradigm Thinking with Fritjof Capra and Brother David Steindl-Rast
These two great minds have joined together to explore the exciting parallels and compatibility of thought in Buddhism, Christianity and modern physics. In the "new paradigm" of thinking, there is a shift from the alienation-producing authority models to one of cooperative thinking, in which we can truly develop a deep sense of belonging to a greater whole. This, they say, is the central aim of spirituality, ecology and perhaps all of human endeavor. Capra is the author of *The Tao of Physics* and Steindl-Rast the author of *A Listening Heart* (Crossroad, 1989); they are also co-authors of *Belonging to the Universe* (Harper San Francisco, 1991).
Tape #2309 1 hr. $9.95

You are a vital part of the work we do!

Please become a member of "Friends of New Dimensions."

We encourage you to become a member of "Friends of New Dimensions" and help bring life-enhancing topics and ideas to the airwaves regularly. As an active member at the individual level or higher, you will receive:

- *New Dimensions* newsletter/journal, a quarterly publication containing feature articles and interviews spotlighting some of the same people and ideas you hear on our radio program, up-to-date program listings for the entire country, descriptions of new tapes, music and book reviews, and items of special interest to New Dimensions listeners.

- A 15% discount on all purchases made through New Dimensions.

Your membership contribution makes it possible to bring life-enhancing topics and ideas to the airwaves regularly, so please join at the level most consistent with your life- or work-style.

Use the order form on the following page. ⇨

New Dimensions Order Form

(U.P.S. cannot deliver to P.O. box)   Date _____

Name _____

Address _____

City _____ State _____ Zip _____

Phone _____

Tape #	Qty.	Title	Amount
1356		Physics and Reality	
1652		The Turning Point	
1685		The Coming Revolution	
1956		Greener Tomorrows	
2094		Quantum Odyssey	
2263		Mindwalk: The New Paradigm	
2309		New Paradigm Thinking	
	1	Tape Catalog	FREE

Check type of payment:
☐ Check or money order ☐ Visa ☐ MC
(payable in U.S. funds)

Acct. # _____

Exp. Date _____

Subtotal	
15% membership discount	
Sales Tax Calif. res. 7.25% BART counties 7.75%	
Shipping & Handling	
Membership	
Total	

Signature—required for all credit card purchases

☐ **YES!** I want to support the radio work and become a member of "Friends of New Dimensions." I understand this entitles me to a 15% discount on all purchases from New Dimensions.

☐ Individual $35 (S721)   ☐ Radio Council: $250 (SP 72)
☐ Family: $45 (S721)   ☐ Satellite Sponsor: $500 (SP59)
☐ Sustaining: $50 (SP94)   ☐ Benefactor: $1000 (SP95)
☐ Radio Underwriter: $100 (S726)

Send order to:
New Dimensions Tapes
P.O. Box 410510
San Francisco, CA 94141-0510
Or order by telephone:
(415) 563-8899
with VISA or MasterCard
ANY TIME, DAY OR NIGHT

SHIPPING & HANDLING

If subtotal falls between	add: U.S. & Canada	Foreign
0-$15.99	$2	$6
$16-$30.99	$4	$8
$31-$50.99	$5	$10
$51-$70.99	$6	$12
$71-$100	$7	$18
over $100	$8	$25

All domestic orders are shipped 1st Class mail or UPS. All orders going outside the U.S. are shipped air. FOREIGN ORDERS: Please send an international bank money order payable in U.S. funds, drawn through a U.S. bank.

OUR GUARANTEE: All New Dimensions tapes are unconditionally guaranteed. If for any reason you are dissatisfied, you may return the tape(s) within 30 days of purchase for a full refund or exchange.

Allow one to three weeks for delivery.   CB

Other Titles in the New Dimensions Books

Marsha Sinetar
in Conversation with Michael Toms
edited by Hal Zina Bennett

Marsha Sinetar is the best-selling author of *Do What You Love, The Money Will Follow* and *Living Happily Ever After.* In her work as an organizational psychologist, she has studied many people who have become successful doing what they love. In this new book, she speaks to those who are attempting to live their deepest calling in the midst of a seductive society. She emphasizes that choosing a lifestyle which blends inner truth with work, family and the demands of twentieth century life is more than possible—it's essential!

$8.95

$8.95

Lynn Andrews
in Conversation with Michael Toms
edited by Hal Zina Bennett

Shamaness Lynn Andrews takes us into the wilderness of self to plumb the depths of our heart so that our being can soar. Her vision quest journey has taken her from the wilds of Manitoba to the jungles of Yucatan and the Aboriginal outback of Australia, as she attempts to bridge the gulf between the primal mind and contemporary life. In this book, she challenges us to see the infinite range of possibilities that lies beyond our ordinary limits—personal and planetary.

Patricia Sun
in Conversation with Michael Toms
edited by Hal Zina Bennett

Patricia Sun is an extraordinary teacher, human energizer and natural healer. In this book she shows how to become more aware of your intuition, and so become more trusting of the Self. With gentle directness, she encourages us to live with spontaneity, continually receptive to the creative force within us. As our words and feelings become aligned with the source of great wisdom within, we assist in the birth of a new world of harmony, cooperation and love.

$8.95

Upcoming Books in the Series:
Larry Dossey in Conversation with Michael Toms
Anne Wilson Schaef in Conversation with Michael Toms

Other Books from Aslan Publishing

Gentle Roads to Survival
by Andre Auw, Ph.D.

This is one of those rare, life-changing books that touches the reader deeply. Drawing from his forty years of counseling as a priest and a psychotherapist, Auw points out the characteristics that distinguish people who are "born survivors" from those who give up, and teaches us how to learn these vital skills. Using case histories and simple, colorful language, Auw gently guides us past our limitations to the place of safety and courage within.

$10.95

The Heart of the Healer
edited by Dawson Church and Dr. Alan Sherr

Bernie Siegel, Larry Dossey, Norman Cousins and sixteen other healing professionals here intimately describe their vision of the healing process and the innermost workings of the true healer. An inspiring and definitive review of the emerging holistic paradigm in healing.

$14.95

Intuition Workout
by Nancy Rosanoff

This is a new and revised edition of the classic text on intuition. Lively and extremely practical, it is a training manual for developing your intuition into a reliable tool that can be called upon at any time—in crisis situations, for everyday problems, and in tricky business, financial, and romantic situations. The author has been taking the mystery out of intuition in her trainings for executives, housewives, artists and others for over ten years.

$10.95

Finding the Great Creative You
by Lynne Garnett, Ph.D.

Finding the Great Creative You is a job and career book for the next phase of the economy. It is focused not on "getting a job" or "climbing the corporate ladder," but doing what you most love as a career path. Practical and down-to-earth, this outstanding book contains dozens of worksheets, exercises and visualizations that allow you to find your deepest purposes—and then actualize them in work.

$10.95

Other Books from Aslan Publishing

Living At the Heart of Creation
by Michael Exeter
Author Michael Exeter is one of the most important voices today for the emerging field of eco-spirituality. *Living At the Heart of Creation* pierces beyond the superficial fixes to the most pressing problems of our day. Blending profound spirituality with wide ecological knowledge, it offers remarkable insights into such challenging areas as the environmental crisis, business, relationships, and personal well-being, inspiring us to live at the heart of creation.

$9.95

Magnificent Addiction
by Philip R. Kavanaugh, M.D.
Kavanaugh's revolutionary work is decisively changing the way we see addictions and emotional disorders. Our unhealthy addictions aren't bad, he says—and it's a waste of time and effort to get wrapped up in getting rid of them, as he demonstrates in his own wrenching personal story. We simply need to upgrade our addictions to ones that serve us better, like addiction to wholeness, life, spontaneity, divinity.

$12.95

Personal Power Cards
by Barbara Gress
An amazing tool for retraining the negative emotions that sabotage most attempts at recovery and personal growth, *Personal Power Cards* work scientifically through colors, shapes and words to re-program the brain for maximum emotional health. Called "One of the most useful recovery tools I have seen" by *New Age Retailer*, these are a simple, incredibly quick and effective technology for building a powerful sense of self-worth in a wide variety of life areas.

$18.95

When You See a Sacred Cow... Milk It for All It's Worth!
by Swami Beyondananda
The "Yogi from Muskogee" is at it again. In this delightful, off-the-wall little book, Swami Beyondananda holds forth on the ozone layer, Porky Pig, Safe Sects, and the theology of Chocolate. Read a few lines and you'll quickly realize that nothing's safe from his pointblank scrutiny.

$9.95

Aslan Publishing Order Form

(Please print legibly)   Date _____

Name _____

Address _____

City _____ State _____ Zip _____

Phone _____

Please send a catalog to my friend:

Name _____

Address _____

City _____ State _____ Zip _____

Item	Qty.	Price	Amount
Marsha Sinetar in Conversation with Michael Toms		$8.95	
Lynn Andrews in Conversation with Michael Toms		$8.95	
Patricia Sun in Conversation with Michael Toms		$8.95	
Gentle Roads to Survival		$10.95	
The Heart of the Healer		$14.95	
Intuition Workout		$10.95	
Finding the Great Creative You		$10.95	
Living At the Heart of Creation		$9.95	
Magnificent Addiction		$12.95	
Personal Power Cards		$18.95	
When You See a Sacred Cow, Milk It…		$9.95	
		Subtotal	
		Calif. res. add 7.5% Tax	
		Shipping	
		Grand Total	

Add for shipping:
Book Rate: $2.50 for first item, $1.00 for ea. add. item.
First Class/UPS: $4.00 for first item, $1.50 ea. add. item.
Canada/Mexico: One-and-a-half times shipping rates.
Overseas: Double shipping rates.

Check type of payment:

☐ Check or money order enclosed
☐ Visa   ☐ MasterCard

Acct. # _____

Exp. Date _____

Signature _____

Send order to:
**Aslan Publishing
PO Box 108
Lower Lake, CA 95457**
or call to order:
(800) 275-2606

NDFC